Igor Tofaneli Tolentino
Ricardo José G. Azevedo

The environmental issue in Belo Horizonte's Master Plan

Igor Tofaneli Tolentino
Ricardo José G. Azevedo

The environmental issue in Belo Horizonte's Master Plan

An analysis of the environment in Belo Horizonte Municipal Law No. 7165/96

ScienciaScripts

Imprint

Any brand names and product names mentioned in this book are subject to trademark, brand or patent protection and are trademarks or registered trademarks of their respective holders. The use of brand names, product names, common names, trade names, product descriptions etc. even without a particular marking in this work is in no way to be construed to mean that such names may be regarded as unrestricted in respect of trademark and brand protection legislation and could thus be used by anyone.

Cover image: www.ingimage.com

This book is a translation from the original published under ISBN 978-613-9-66150-3.

Publisher:
Sciencia Scripts
is a trademark of
Dodo Books Indian Ocean Ltd. and OmniScriptum S.R.L publishing group

120 High Road, East Finchley, London, N2 9ED, United Kingdom
Str. Armeneasca 28/1, office 1, Chisinau MD-2012, Republic of Moldova, Europe
Printed at: see last page
ISBN: 978-620-8-01347-9

SUMMARY

The environmental issue is becoming increasingly important in the contemporary world, due to the environmental impacts of urban-industrial society. In this context, urban planning, with the Master Plan as an essential instrument, is important for territorial planning and improving the population's quality of life through the sustainable development of the urban environment. The aim of this work is to analyse how the environmental issue is dealt with in Belo Horizonte's current Master Plan, by means of a qualitative analysis that takes into account the guidelines related to the environment, outlined by urban planning. The methodological procedures include a bibliographical survey on urban planning and the environment, a study of municipal legislation related to the subject, as well as interviews with public managers and social agents involved with the environmental issue in Belo Horizonte. It is hoped that the knowledge obtained from this research will subsidise the development of new work related to the urban environment of the capital of Minas Gerais, as well as promoting public policies aimed at a satisfactory quality of life for the population.

Keywords: Urban environment; Urban planning; Masterplan.

Field of Expertise: 7.06.01.03-8 - Urban Geography

SUMMARY

CHAPTER 1

INTRODUCTION

The city has its origins at a certain time in human history and is founded over the course of the historical process, taking on various forms and contents, as Carlos (2013) shows in his book **The City**. The historical origin of cities was around 5000 BC, with the formation of small settlements near rivers such as the Euphrates. In the Middle and Modern Ages, the city began to take on a new form, especially with Feudalism and the Crusades, in which land and trade became a source of wealth, as well as absolutism. In the Contemporary Age, beginning with the French Revolution (1789), capitalism develops and consolidates, there is the Technical-Scientific Revolution, as well as industrial development, which creates a new spatialisation and a new division of labour. This brings us to today's capitalist city, which is completely different from the medieval city.

Chaos. Movement. Cars. Congestion. Noise. Violence. Crime. Streets. Avenues. Houses. Buildings. Shopping centres. Shopping centres. Centre. Crowds. Time. Hurry. Nowadays, these are some of the words and sensations immediately associated with the city by the majority of its inhabitants, especially in large metropolises and, if it is only interpreted in this way, the city ends up acquiring a reified meaning[1] , because it doesn't go beyond these perceptions. These refer to what would be the portrait of the urban landscape, that is, what is instantly perceptible in space, the dimension of space that can be intuited, which here is urban space.

The city, however, comprises more than just the built and the chaotic. When you look at it from an architectural point of view, through the distinctive shapes and characteristics of the building that portray a particular period in history, what you see

---

[1] Reified here means the thingification of the city, its materialisation, that is, the transformation of the abstract concept of the city into a concrete reality.

are several cities within a single one. However, behind this static appearance lies the movement of life and, when you look at a particular place, it reveals a specific moment in the daily lives of the people who live, work and move around the city (CARLOS, 2013). It reveals "all the dynamism of the landscape's process of existence, the product of a relationship based on contradictions, in which the pace of change is given by the pace of development of social relations" **(CARLOS, 2013, p.38)**. In this way, the urban landscape not only reproduces the historical process, but is also a product of it (CARLOS, 2013).

In this context, urban space is not static, but dynamic. It is built through an unequal process, determined by the characteristics of the process of capital reproduction and accumulation, but it also refers to the development of human life in all its dimensions and meanings, and is determined by the reproduction of social relations. Thus, urban space is the historical product, at a given moment, of the state of society; it is the result of the activity and dynamism of society which, over time, has acted on it, transforming it, humanising it, making it a product increasingly distant from the natural environment (CARLOS, 2013).

For Carlos (2013, p. 33) "space is a social product in an uninterrupted process of reproduction", a concept that Milton Santos describes as the roughness of space:

> Let's call roughness what remains of the past as form, built space, landscape, what remains of the process of suppression, accumulation, superimposition, with which things are replaced and accumulated everywhere. Roughness presents itself as isolated forms or as arrangements (SANTOS, 2012, p. 140).

Capitalist urban space also refers to the integration of different land uses in a city, defining areas (industrial, residential, commercial, etc.), and is therefore the spatial organisation of the city; it is both divided and connected, as its parts relate to each other empirically through the flow of people and cars, the circulation of goods and capital, etc. The core of these connections has traditionally been the city centre. According to Corrêa (2005, p. 9), this capitalist urban space is "fragmented and articulated, a reflection and social conditioner, a set of symbols and a field of class struggles; [...] it is the result of actions engendered by agents who produce and consume space".

The relationship between urban space and society is presented in different ways at different levels of development. Firstly, the city appears as a form of appropriation of this produced space, expressed through the use of urban land. In capitalist society, the way in which this use takes place will be through the process of exchange, i.e. appropriation, which takes place on the property market, via private property, arising from the need to produce, consume, live or dwell. The value of urban land depends on various factors and is expressed through its location.

The city, then, is seen as a product to be consumed according to the laws of capital reproduction. The way in which land is occupied will be determined by the different segments of society in different ways, generating conflicts between individuals and uses, which will be guided by the market, which produces a limited set of choices and living conditions (CARLOS, 2013). Space thus appears as materialised social work, appropriated differently by citizens.

The urban landscape of the capitalist city then shows a certain heterogeneity between ways of life, ways of living and, above all, the use and occupation of land. What can be seen is spatial segregation, a product of social inequality. The contrast is very visible between noble neighbourhoods and working class neighbourhoods. Still in the capitalist city, especially in large metropolises, the urban landscape is an expression of "order and chaos", where time "is money" and the traffic light ends up being a symbol of time and the intense rhythm of these places. Relationships between people end up being measured by merchandise, by money, and citizens become more insensitive, individualistic and repugnant (CARLOS, 2013).

Society is hierarchical, divided into classes. The higher-income classes, who used to live in the city centres, now increasingly seek to live in the areas around the cities, in new neighbourhoods or gated communities with more amenities, including greater tranquillity, a better quality of life, greater proximity to the natural environment and greater security, because, according to Corrêa (2005, p. 71), "residential use based on a low-status population and the physical deterioration of the central area have stigmatised it, creating an image of poverty, vice and crime". The lower-income class then has the possibility of housing in the tenements located close to the city centre, the housing estates produced by the state, as well as the favelas, on land that has often been invaded, in which these excluded social groups become shaping agents, producing their own space (CORRÊA, 2005). These are often in locations with fewer amenities and precarious housing and living conditions.

In this sense, the state creates the conditions for the reproduction of capitalist society, the social classes and their fractions, managing conflicts that could interfere with the realisation of the capital cycle, whether by producing infrastructure, controlling wages to keep them low, etc. This tends to favour the classes and segments of society that have a higher status or are in power. The capitalist state then mystifies and softens the contradictions of the system, imposing on the life of society the needs of the dynamics of capital accumulation (CORRÊA, 2005; CARLOS, 2013).

Based on this and the urban contradictions generated by capitalist society, urban social movements have emerged to fight for the right to the city in its fullest sense, as well as questioning the standardisation of the city, urban life and the actions of the state. Thus, the city is seen as the stage for the class struggle. According to Carlos (2013, p. 87), "more than a right to the city, what is at stake is the right to obtain from society those minimum goods and services without which existence is not dignified. It is the right to participate in a society of excluded people".

At the heart of the urban crisis is the power seen in private land ownership, which gives rise to the current standardisation of access to the city, both in terms of housing and living conditions, expressed in the contradiction between wealth and poverty. Capitalism's inability to meet the needs of an increasingly large portion of the population is evident (CARLOS, 2013).

Today's urban society presents us with a changing scenario, with a tendency towards the dissolution of social relations, the destruction of social memory and a profound change in the urban way of life. "In the spectacle of the crowd, the individual is lost and, for him, the city becomes either a passageway or a shop window," writes Olgária Matos (1998, p.16) in The City and Time. Carlos (2013, p.91) writes that "there are residues and resistances in the underground that escape the homogenising and terrifying process of capital".

It is necessary to think about a more humane city, a new urban, which means overcoming the current capitalist economic, social, political, ideological and legal order, through the participation of the whole of Brazilian society, prioritising democratic conquests, in order to advance towards the construction of a new society. In this way, this work aims to analyse how the environmental issue is addressed in Belo Horizonte's Master Plan, one of the country's most important capitals, as well as establishing a relationship and presenting the importance of building an urban space, from its planning stage, that is integrated with environmental issues and perceptions.

CHAPTER 2

URBAN PLANNING

When we talk about planning, we intuitively think of something in the future, it means designing or architecting something, trying to predict the evolution of a phenomenon in order to adapt so that its objectives can be achieved. According to Souza (2013, p. 46), planning means "trying to simulate the unfolding of a process, with the aim of better guarding against probable problems or, conversely, with the aim of taking advantage of probable benefits". However, before talking about the purpose of urban planning, we need to talk about urban management. Managing relates to something in the present, and means "administering a situation within the framework of the resources currently available and with a view to immediate needs" (SOUZA, 2013, p. 46).

In this way, urban planning is the preparation for future management, aiming to minimise problems and increase room for manoeuvre (SOUZA, 2013). Even so, according to Duarte (2007), planning

> [...] recognises, locates, trends or natural propensities (local and **regional) for development, as well as "establishes the rules of** land **occupation,** defines the main strategies and policies of the municipality and spells out the restrictions, prohibitions and limitations that must be observed in order to maintain and **increase the quality of life for its citizens"**[2] . (p. 22)

The issue of urban planning in Brazil has undergone changes since 1895, when planning focused on the development of parts of cities and did not involve the city as a whole, until the late 1980s and the present day. The latter have seen a return to democracy, greater popular mobilisation and, consequently, the politicisation and institutionalisation of planning (RODRIGUES, 2013).

Another important fact in relation to the issue, and one that gave it a new impetus

---

[2] REZENDE, D.; CASTOR, B. V. J. *Planejamento estratégico municipal*. Rio de Janeiro: Brasport, 2006.

in the 2000s, was the approval of the City Statute in 2001, which regulates articles 182 and 183 of the Federal Constitution of 1998 and "establishes rules of public order and social interest that regulate the use of urban property in favour of the collective good, the safety and well-being of citizens, as well as environmental balance" (Law no. 10.257, 2001). Chapter III of the Statute deals with the masterplan, which is the main instrument for guiding the development of urban space (RODRIGUES, 2013). In addition, the Ministry of Cities was created in 2003, demonstrating the importance of discussing urban policy and emphasising the right to the city for its inhabitants through the principle of the social function of property.

Creating a plan to optimise the achievement of urban planning objectives is necessary and, moreover, mandatory, according to the provisions of Law No. 10,257, in its third chapter. This plan is the Master Plan, which, according to article 40 of the City Statute, is **"the basic instrument of urban development and expansion policy" and "an** integral **part of** the municipal planning process, with the multiannual plan, budget guidelines and annual budget incorporating the guidelines and priorities **contained** therein". **Also according to Villaça and Silva, the Master Plan**

> It would be a plan that, based on a scientific diagnosis of the physical, social, economic, political and administrative reality of the city, the municipality and its region, would present a set of proposals for future socio-economic development and the future spatial organisation of urban land uses, infrastructure networks and fundamental elements of the urban structure, for the city and the municipality, proposals defined for the short, medium and long term, and approved by municipal law. (VILLAÇA, 1999, p. 238)

> It is a plan, because it establishes the objectives to be achieved, the timeframe in which these are to **be** achieved [...], **the activities to be carried out and who is to** carry them out. It's a masterplan because it sets the guidelines for the municipality's urban development. (SILVA, 1995, p. 124 - emphasis in original)

A city's Master Plan is the responsibility of all citizens and everyone has the right to participate in public hearings. It is also important for the population to ensure that it reflects their wishes for the city and that it is of good quality. The Plan is drawn up in stages and the basic guidelines of the Plan are proposed by a multidisciplinary technical team, including engineers, geographers, urban planners, economists, health professionals, politicians, among others, to make the plan better structured (DUARTE, 2007). Once the plan has been drawn up, it must be passed into law by the city council. Once approved, some aspects need legal, financial and technical complementation for it to be implemented. In addition, it needs to be managed by the municipal government so that it can be updated according to the evolution and changes taking place in the city, and so that its recommendations are respected (DUARTE, 2007).

In addition to the Master Plan, there is another very important aspect to the process of carrying out urban planning, so that it reflects and comes closer to the reality of the city: environmental perception. Firstly, environmental perception is a mental activity that takes place through the processes of cognition and perception, involving the senses of the human being in an activity of interaction between them and the environment; it is where the individual will establish a link between themselves and the space around them, be it affective or rejection, in other words, topophilia or topophobia (ABRANCHES, 2009).

Studies on environmental perception are fundamental to understanding man's inter-relations with his space, whether individually or socially, by analysing his expectations, judgements, behaviours and limitations. In other words, how the individual or community interprets, sees and acts in relation to their environment according to their interests, needs and desires, receiving influences from the elements that make up their cultural heritage, such as values and customs (ABRANCHES, 2009).

> In planning processes, [...] these studies are fundamental because they provide knowledge of the particularities of each society/individual/environment relationship, making it possible to develop programmes that really promote participation and solutions in line with social demands. (ABRANCHES, 2009, p. 498)

Urban planning therefore has the city as its object of intervention, and this must be understood as a product of spatial processes, "as a complex, unpredictable phenomenon, the result of diverse interests and built by planners representing public authorities and civil society" (ABRANCHES, 2009, p. 500). Within the city there are several regions with varying factors, such as the level of education of its population, income level, lifestyle, among others. From this, it can be seen that environmental perception is an essential aspect of urban planning, especially in relation to the production of a city's Master Plan, as this will provide more reliable and legitimate data for future proposals to improve the environmental quality (ABRANCHES, 2009) and quality of life in the municipality.

In this way, urban planning must reflect the characteristics and needs of the city, promoting the participation of all social actors in the decision-making process, and bringing together knowledge and professionals from different areas to formulate guidelines, policies and strategies for future changes in the urban space, as well as developing programmes defined according to local identity. These aspects will therefore help make public management more effective and efficient for the city, enabling a better development and future for the city and bringing it closer to what is idealised by its citizens.

2.1 Urban planning in Belo Horizonte

The city of Belo Horizonte began to be planned in 1894 by the New Capital Construction Commission, with Aarão Reis as chief engineer. The new capital was to be built in the Arraial de Belo Horizonte, formerly known as Curral Del Rey, which was founded in 1701 by the bandeirante João Leite da Silva Ortiz.

The Commission's process of producing the new city plan was strongly influenced by the urban reforms in Paris carried out by Baron Georges-Eugène Haussmann, when he was mayor of the French capital between 1853 and 1870, especially in relation to concerns about the health of the new capital's population and aesthetics (COSTA; ARGUELHES, 2008). The plan was delivered to the government of the state of Minas Gerais on 23 March 1895, by means of Official Letter No. 26 (COSTA; ARGUELHES, 2008) and the capital was inaugurated on 12 December 1897 by President Chrispim Jacques Bias Fortes, under the name of Cidade de Minas, and would be renamed Belo Horizonte four years later.

According to the urban plan, the municipality was divided into three zones - urban, suburban and rural (sites) - with the urban zone having a much more detailed infrastructure, with the aim of receiving civil servants and the government's administrative-bureaucratic apparatus from the former capital, Ouro Preto. In addition, it was planned to be occupied by a population of 200,000 inhabitants and, as idealised by the Commission, this occupation would take place from the centre to the periphery (COSTA; ARGUELHES, 2008). With regard to environmental issues, it can be seen that there was great concern about green areas and landscaping, as the plan proposed a large park in the central area and, in addition, the avenues would have trees on the sides and the streets would be lined with trees in the middle.

Over time, Belo Horizonte grew in a disorganised way, contrary to what was envisaged in its initial urban plan. In the 1930s, the city went through an urban crisis due to accelerated growth.

> Belo Horizonte's growth was then mainly due to the addition of new allotments in peripheral areas. Real estate speculation took on alarming proportions and worried the local government due to the difficulties of urbanisation imposed by this type of occupation. (MELO, 1991, p. 42)

Furthermore, the production of urban space in Belo Horizonte did not take into account the social demands of the less favoured groups and the suburban area was occupied in a disorderly fashion.

> **In turn, the suburban area was occupied in a "disorganised" way and without** any concrete attention from the public authorities. As time went by, the process of socio-spatial segregation characteristic of the city's layout intensified, as did the government's disregard for the less favoured classes, albeit disguised by countless decrees and laws (COSTA; ARGUELHES, 2008, p. 130).

Faced with this disorganised growth, some measures were taken during the administration of mayor José Soares de Mattos (1933-1935), such as the creation of the Technical Advisory Commission for the City in 1934, which was made up of Luiz Signorelli, Ângelo Murgel, Fábio Vieira, both architects, and engineer Lincoln Continentino (COSTA, 2006; BAHIA, 2005). This new commission had the responsibility of studying a regulatory plan for the situation the city found itself in (BAHIA, 2005), i.e. to carry out a new plan for the city's development. Continentino submitted a proposal for the capital's expansion, which had not been fully implemented, but included aspects such as

> [...] the establishment of a system of large avenues, linking the urban area to the suburbs and neighbouring towns; the unification of the railways in operation in the city; the replacement of streets with others more adapted to the topography; zoning with the division of the city into three zones, residential, commercial and industrial; and the drawing up of a building code for the city. (COSTA, 2006, p. 6)

During his administration (1940-1945), Juscelino Kubitscheck made several urban and architectural changes to the capital of Minas Gerais, such as the construction of the Pampulha and Cidade Jardim neighbourhoods, the Industrial City, the University City, the extension of Avenida Amazonas and the opening of Avenida Presidente Antônio Carlos, among others (COSTA, 2006; BAHIA, 2005). At the end of the 1940s, the then mayor (1947-1951) Otacílio Negrão de Lima sent a proposal for spatial reorganisation to the city council, by creating satellite cities around the capital, with different functions: "Barreiro would have an agricultural function, Cidade Industrial a manufacturing centre, Venda Nova a residential centre, and Pampulha an entertainment centre" (COSTA, 2006; p. 7). This was justified because Belo Horizonte was unable to provide adequate infrastructure for the growing population (BAHIA, 2005).

In the 1950s (Figure 1), Belo Horizonte underwent a developmental and modernist impulse, with the supply of electricity and the development of a transport network, as well as the creation of companies such as Cemig, Usiminas and others, and the expansion of water supply services (COSTA, 2006; BAHIA, 2005), enabling the capital to establish itself as an urban-industrial centre. Since then, the city has grown and developed, and urban planning has become increasingly necessary to try to keep up with the reality of the capital of Minas Gerais and help it evolve.

During his time as mayor (1951-1954), Américo Renné Giannetti, concerned about city planning, created the Master Plan Service, calling on Francisco Prestes Maia, Oscar Niemeyer and Burle Marx to draw it up. The plan had to be rational and reflect the reality of the city and the socio-economic conditions of its inhabitants (COSTA, 2006; BAHIA, 2005), and was thus described by Giannetti in 1952:

> The creation of the master plan service, as pointed out in the Administration Programme Plan, was an unavoidable measure to re-establish the continuity of the planning established above all by the Capital Construction Commission. The absence of a service of this nature in the administrative organisation of the City Hall over a period of more than 40 years, where the responsibility for the disordered expansion in the suburban area, of neighbourhoods and towns that sprang up without obeying an oriented urban plan (...) to plan new housing nuclei (...) the preparation of a complete and precise Cadastral Plan and the prior study of the conditions of the environment, historical, social and economic factors and existing legislation, tasks that were immediately faced by the municipal government. (PLAMBEL, 1979, p. 299)

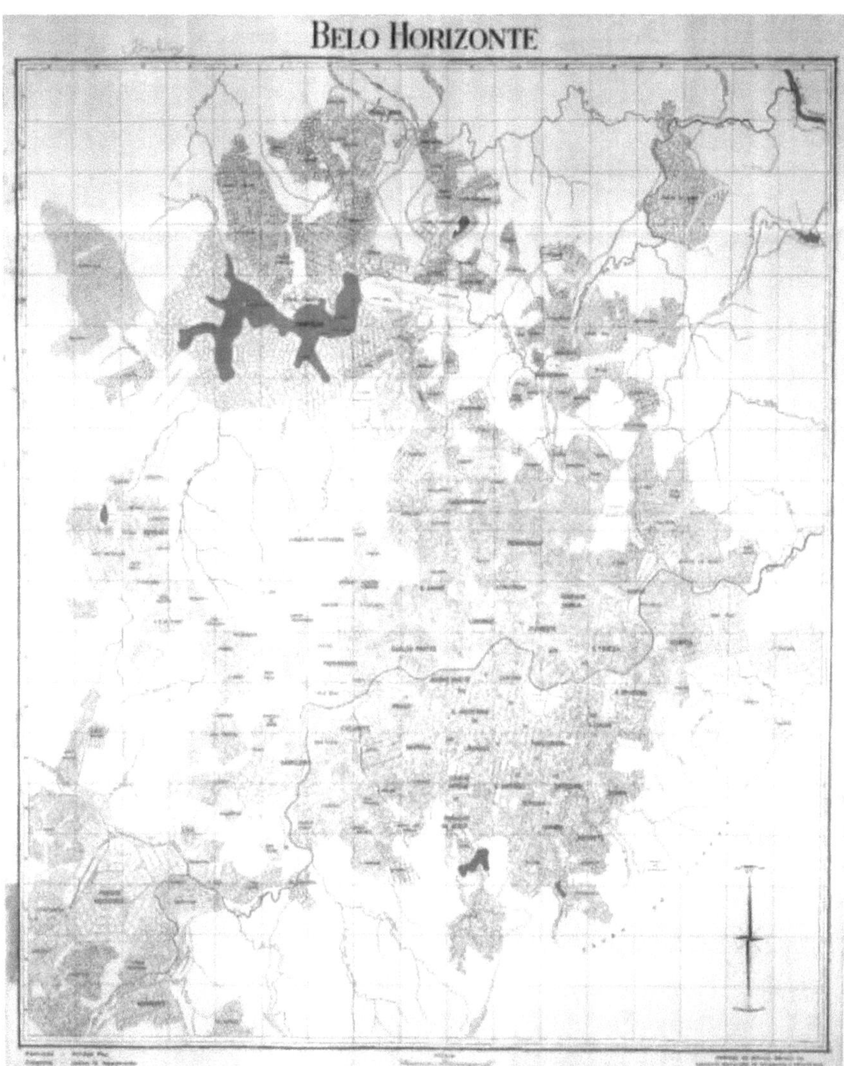

Figure 1 - Map of Belo Horizonte (1953). The urban expansion towards the west and north can be seen, driven by the Industrial City and Pampulha, respectively. Source: PANORAMA de Belo Horizonte; Historical Atlas. Belo Horizonte, João Pinheiro Foundation; 1997.

However, according to Costa (2006), there is no news of a proposed masterplan ever materialising. Giannetti's successor, Celso Mello de Azevedo (19551959), made a more significant contribution to the issue of urban space in Belo Horizonte (COSTA, 2006). During his administration, a contract was signed with the Society of Graphic and Mechanographic Analyses Applied to Social Complexes (SAGMACS) to carry out a detailed analysis of the capital's urban space (BAHIA, 2005). The study found a serious situation regarding the quality of life of the capital's population, 47 per cent of whom were living in conditions considered subhuman, and noted that there was a lack of infrastructure and urban facilities, as well as a high rate of property speculation (MELO, 1991; PLAMBEL, 1979). With regard to the Masterplan, it

> [...] it proposed the decentralisation of the commerce and services sector, through the establishment of sub-centres boosted by the hierarchisation of the road system and the creation of "commercial corridors". In relation to property speculation, measures were proposed such as regulating new subdivisions, progressive taxation, inspection and zoning that would support more efficient action by the public authorities. (COSTA, 2006, p. 8)

The Plan would serve as a reference for subsequent ones, although it was not implemented (COSTA, 2006).

By the end of the 1950s, urban interference and disorder were already noticeable in the municipality, because the municipal government was unable to respond to this accelerated metropolisation and its modernisation was still a process of economic development based on industrialisation (BAHIA, 2005). Furthermore, in 1964, the year of the beginning of the military dictatorship in Brazil, it was at federal level that the power of the state became stronger, **which "guided** an economic model defined by an economy of income concentration, backed by the hegemony of foreign capital, causing social inequality in spatial and sectoral terms" **(BAHIA, 2005, p. 197).** In this way, the concepts and process of modernising the capital were compromised, essentially in terms of social aspects, pushing the municipality into the background (BAHIA, 2005).

Currently, Belo Horizonte has a resident population of 2,375,151 inhabitants and a territorial unit area of 331.401km² , covering nine managerial subdivisions of the municipality: Barreiro, Centro-Sul, Leste, Nordeste, Noroeste, Norte, Oeste, Pampulha and Venda Nova (PBH, 2008). According to information on the Belo Horizonte City Hall website, the division of the municipality into administrative regions is essential for decentralisation, as well as for the implementation and effectiveness of programmes and activities that meet the specific needs of each region.

2.1.1 The 1996 Master Plan

In 1985, with the establishment of the New Republic, proposals for more democratic forms of planning with greater involvement of the population were increasing (COSTA, 2006). In the 1990s, the issue of functionalist urban planning in Belo Horizonte was characterised **by a break with "the Master Plan** approved in 1996 clearly representing recent trends in **urban** planning" **(COSTA,** 2006, p. 12).

Belo Horizonte's Master Plan, Municipal Law No. 7165/96, was drawn up with popular participation through public hearings, although in practice this participation was restricted (COSTA, 2006). Its aim is to organise territorial planning, the sustained development and functioning of the city, guide the actions of public authorities and private initiative, and prioritise public investments. Thus, the Plan defines the strategic objectives of urban development, guidelines for urban policy, macro-zoning and its instruments, as well as an urban management model to be implemented (CÂMARA MUNICIPAL).

The 1996 Plan is organised into titles, chapters, sections and subsections, and is set out in articles, since it is a law. It contains general provisions, concepts, global and strategic objectives, urban policy instruments, and brings together general and specific guidelines that are essential for urban policy and must be taken into account when drawing up various municipal public policies, such as: economic development, public intervention in the urban structure - which includes environmental, sanitation, educational, housing, etc. guidelines - social and zoning guidelines, among others. It also establishes areas of special guidelines, social interest and how citizens will participate in urban policy.

With regard to the social function of property, according to Law No. 7.165/96, this materialises when the property complies with the land-use planning criteria and urban development guidelines contained in that same law, ensuring, as stated on the Belo Horizonte City Council's website: "the socially just and rational use" of urban space; "the appropriate use of available natural resources, as well as the protection, preservation and recovery of the environment"; "the use and utilisation compatible with the safety and health of users and neighbours". **Although it was approved and implemented in 1996, in the** following **years** the Master Plan underwent several modifications, examples being Law No. 8.137/2000 and Law No. 9.959/2010, which have already been added to the original text of Law No. 7.165/96.

In addition, in the same year that the Plan was implemented, the guidelines for urban development were put into effect through the Urban Land Use, Occupation and Parceling Law, Municipal Law 7166/96. As Costa explains, this law

> [...] sought to be more flexible than previous laws with regard to the location of non-residential activities, in an attempt to adapt to the diversity of the urban environment. To this end, a macro-zoning of occupation was established, with broader criteria related to infrastructure, topography, the road system, accessibility and the demands of environmental preservation and protection in the various regions of the city. Non-residential uses were initially made more flexible throughout the municipality, with installation permitted according to the classification of the activity, its nature and the width of the public road. (COSTA, 2005, p. 13)

With regard to the environmental issue in the Belo Horizonte Master Plan, it can be seen that there is considerable importance given to the subject, with a specific subsection within Section II of Chapter III, The Guidelines for **Public** Intervention **in the Urban Structure, entitled "Subsection IX, The Environment". The subsection is** made up of 2 articles, which describe the concept of the environment and establish the Municipal Environmental Policy, as well as setting out the 37 guidelines relating to the subject.

CHAPTER 3

METHODOLOGY

The methodology used to achieve the objectives of this project followed the schedule of activities presented in the proposal submitted to Call for Proposals N°184/14, which deals with the Call for Proposals for the Institutional Junior Scientific Initiation Scholarship Programme (BIC-Jr.), as shown in Table 1. It should be noted that this timetable is based on seven stages, so that the fulfilment of each of them was essential for the development of the research, as well as for the production of this final report.

Chart 1: Timetable of activities

Activity	Month											
	1	2	3	4	5	6	7	8	9	10	11	12
1 - Bibliographical survey.	X	X	X	X								
2 - Research into environmental legislation, categorisation and analysis of the environmental guidelines in the Master Plan.				X	X							
3 - Description of the main actions and public policies aimed at the environment in the municipality.				X	X	X						
4 - Qualitative analysis of environmental legislation and policies.					X	X	X					
5 - Interviews with municipal public managers involved in environmental issues / submission of a letter to the City Council.							X	X				
6 - Preparation of a report: analysing the environmental issue in the Belo Horizonte Master Plan.								X	X	X	X	
7 - Delivery of the final report.												X

Firstly, several bibliographies were consulted that dealt with the content that would be covered in the research, especially on urban planning in general and specifically in Belo Horizonte, including reading and summarising books, articles and other academic works, as well as the Belo Horizonte Master Plan itself, Law No. 7.165/96.

After the bibliographical survey, the environmental guidelines in the Plan were categorised and analysed. This phase also included researching and describing municipal legislation and public policies related to environmental issues.

In the next stage, a qualitative and quantitative analysis was made of the legislation found, but above all of the public policies or environmental programmes implemented in the municipality, in order to find out which of the guidelines have not yet been implemented in the municipality. The sources used were articles, laws, news and various other types of texts containing the information needed for the research, found on the websites of municipal bodies or other reliable sites. Graphs and tables were also drawn up with some of the data obtained for better visualisation, organisation and display.

In the next phase, the authors of the research tried to conduct an interview with a public authority involved with environmental issues in the municipality, but it wasn't possible. Based on the discovery of guidelines that have not yet been implemented and some doubts that arose during the development of the research, a letter was sent to Belo Horizonte City Hall containing questions on the subject to be answered by public authorities responsible for or involved with the issue addressed in the project. However, the authors of the research have not yet received a response from the city council.

Finally, this final report was drawn up, containing all the information and analyses carried out during the research, including the text produced with the bibliographical survey, which is of essential importance for understanding the subject of the research.

CHAPTER 4

RESULTS AND DISCUSSION

The environmental guidelines in the Master Plan (PD) were analysed and grouped into categories to better identify the environmental themes addressed in the PD and their corresponding guidelines, as shown in Table 1 and Figure 2 below. It can be seen that most of the 37 environmental guidelines relate to green areas and environmental control, and only one relates to the use of sustainable technologies. Furthermore, it can be seen that the public authorities' view of the environment is strongly linked to its physical-ecological dimension, which is dissociated from the social dimension that is closely related to urban environmental problems.

Table 1: Categories of environmental guidelines

	Categories						
	Green Areas	Environmental Control	Water Resources	Geomorphology	Environmental education	Information System	Sustainable Technologies
Number of guidelines on environmental issues, Art. 22 of the Master Plan (Law No. 7.165/96)	I, II, III, IV, V, VII, IX, XIV, XV, XVI, XVIII, XXII, XXX, XXXI, XXXII, XXXIII, XXXVI	IX, X, XI, XII, XVII, XIX, XX, XXI, XXIV, XXV, XVII, XXVIII, XXIX, XXXIV, XXXV, XXXVII	IV, VI, VII, VIII, XXI	VII, VIII, XIII	XXIII, XXIV, XXVII	XX, XXIII, XXVI	XXXIV

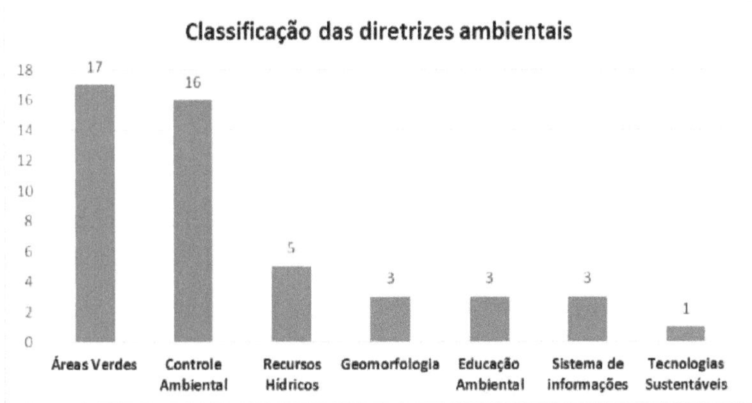

Figure 2 - Graph illustrating the quantification of environmental guidelines in the Belo Horizonte Master Plan, by thematic class.

However, it is worth noting that, despite the categorisation, the guidelines can be interrelated, i.e. the same guideline can present aspects of more than one environmental thematic class. And if we look more closely at what each of these classes is and relate them to the guidelines, we can see that:

- Green areas: the guidelines deal with the creation, maintenance, preservation, recovery of municipal green areas, restoration of riparian forests, afforestation of public places, as well as the implementation of new areas, among other related specificities;

- Environmental control: the guidelines mention the creation of pollution control mechanisms, the recovery of degraded areas, solid waste management and the development of public policies, technical standards and legislation for activities that cause an environmental impact, among others;

- Water resources: the guidelines express the promotion of the recovery and protection of municipal water bodies, the increase in soil permeability rates, the containment of silting processes, as well as guiding the drafting of legislation on the use and control of groundwater;

- Geomorphology: guidelines deal with the stabilisation of slopes at risk of landslides, the containment of earth movements and the decapping of the soil and increasing its permeability;

- Environmental education: the guidelines state that environmental education should be prioritised through the media, in addition to the implementation of projects, programmes and activities in various locations, as well as the promotion of citizen participation on environmental issues;

- Information system: guidelines mention the creation of a system containing information on the environment, for public access and integration between the municipality's environmental bodies and state and federal entities;

- Sustainable technologies: the guideline only expresses the encouragement and adoption of sustainable technologies in actions developed by the public and private sectors.

After categorising the DP's environmental guidelines and researching the environmental legislation and public policies implemented in the municipality, the laws and/or public policies were related to the environmental guidelines.

In this way, it was possible to find out which guidelines were or were not covered by municipal environmental laws and/or public policies, or that in some way these were related to what was covered in the guidelines, as shown in Table 2 below, where the "X" in red means that they were not covered, i.e. no specific legislation and/or public policies were found.

It also shows some municipal public bodies that can carry out functions that are expressed in the guidelines. In addition to this table, two other tables were produced, one showing the description of the guidelines (Table 3), and the other showing the description of the legislation (Table 4) mentioned in Table 2.

Table 2: Contemplation of environmental guidelines through legislation and/or municipal environmental public policies

Guidelines	Legislation	Public policies/responsible bodies
1	X	Within the Municipal Environment Secretariat, there is the Environmental Management Department (GGAM) and, subordinate to it, the Green Areas and Urban Tree Planting Department (GEAVA).
2	X	Adopt Green, Tree Inventory, Garden City Competition, Greener BH.
3	X	X

4	X	X
5	Law No. 6.314/93	X
6	X	PROPAM - Programme for the Recovery and Environmental Development of the Pampulha Basin - public policy applied only to the Pampulha Lagoon.
7	Law No. 7.166/96	X
8	X	X
9	Law 8.616/2003 and Law 10.522/2012	Sustainable Management System for Construction Waste and Voluminous Waste (SGRCC) and the Municipal Plan for Integrated Management of Construction Waste and Voluminous Waste (PMRCC)
10	Law^0 4.998/88	X
11	Law^0 7.277/97	X
12	X	Integrated Development Master Plan for the Belo Horizonte Metropolitan Region - there are very relevant aspects related to the environmental issue, such as policies and programmes, which can be seen in the PDDI's final report, available on the Internet.
13	X	PEAR - Structural Programme in Risk Areas. The programme is coordinated by URBEL - Companhia Urbanizadora e de Habitação de Belo Horizonte.
14	Law No. 9.011/2005, Decree No. 14.370/2011 and Decree No. 14.708/2011	Maintenance is carried out by the Municipal Parks Foundation. Adopt Green Programme.
15	X	BH Mais Verde, Private Ecological Reserves, Adopt the Green, Tree Inventory, Garden City Competition.
16	X	Greener BH.
17	Law No. 9.505/2008, Law No. 8.262/2001 and Law No. 10.175/2011	Municipal Policy for Mitigating the Effects of Climate Change, Municipal Policy for Mitigating the Effects of Climate Change and Municipal Plan for Reducing Greenhouse Gas Emissions.
18	Decree 14.060/2010	X
19	X	X
20	X	PEAR, Operation Oxygen, other related

		programmes.
21	X	X
22	X	X
23	X	Environmental Education Extension Centre / Green Room (CEEA) and Regional Environmental Education Centres (CEAs)
24	Decree No. 15.745/2014	Superintendência de Limpeza Urbana (SLU) - responsible for solid waste management. There is also the Tyre Reception Unit, Small Volume Reception Units and the Composting Programme.
25	Decree 5.893/88 and Law^0 4.253/85	Policy for the protection, control and conservation of the environment and improvement of the quality of life in the Municipality of Belo Horizonte and Municipal Environmental Policy
26	X	No specific information system was found, apart from the Belo Horizonte City Hall website, in the part of the Municipal Environment Secretariat.
27	X	Programmes such as Adopt the Green and the Garden City Competition, among others.
28	Law^0 9.505/08	X
29	Law No. 4.495/1986, Law No.^{0} 9.340/2007 and Law No.^{0} 10.175	Operation Oxygen Programme, Belo Horizonte Biodiesel Programme and the Municipal Policy for Mitigating the Effects of Climate Change.
30	X	X
31	X	X
32	Law No. 6.314/93	X
33	X	X
34	X	X
35	X	X
36	X	X
37	Decree No. 5.893/1988 and Law No. 4.253/1985	Policy to protect, control and conserve the environment and improve the quality of life in the Municipality of Belo Horizonte

Table 3: Description of the 37 environmental guidelines in the Masterplan

Guidelines	Description of the guideline
1	Delimit appropriate spaces that have the characteristics and potential to become green areas;

2	Enabling the afforestation of public places, especially in regions lacking green areas;
3	Delimit areas for the preservation of ecosystems;
4	Delimit *non-aedificandi* strips to **protect water banks and springs,** to maintain and recover riparian forests;
5	To guarantee the preservation of vegetation cover of environmental interest in private areas, by means of compensation mechanisms for owners;
6	Promote the recovery and preservation of municipal lakes, reservoirs and lagoons;
7	Ensuring higher levels of soil permeabilisation in public and private areas;
8	Control soil stripping and earth movements in order to avoid silting up dams, streams, reservoirs and lagoons;
9	Draw up urban plans for landfill sites, preferably using them to recover degraded areas and subsequently create green areas;
10	Establish criteria for the installation and control of activities that involve safety risks, radioactivity or that emit pollutants, vibrations or radiation, implementing an effective and up-to-date inspection system, especially in places where X-ray machines are used;
11	Define and regulate, in specific legislation, the works and activities that cause an environmental impact, in relation to which special licensing procedures must be adopted;
12	Promote coordination with the municipalities of the Metropolitan Region to develop urban planning programmes of common interest, through environmental control mechanisms, technical standards and compensation for damage caused by pollution and environmental degradation;
13	Promote the stabilisation of slopes at risk of landslides;
14	Recover and maintain green areas, creating new parks and squares;
15	Ensure the proportion of at least $12m^2$ (twelve square metres) of green area per citizen, distributed by regional administration;
16	Prioritise the creation of green areas in regional administrations where the rate does not reach that set out in the previous point;
17	Establishing effective control of noise, visual, atmospheric, water and soil pollution, setting quality standards and monitoring programmes, especially in critical areas, with a view to their environmental recovery;
18	Establish a programme to create conditions for the survival of birds in urban areas by planting fruit trees, under the terms of Federal Law No. 7,563 of 19 December 1986;
19	Require mining companies to recover degraded areas;
20	Establishing the integration of municipal environmental bodies with state and federal environmental control entities and bodies, with a

	view to increasing effective joint actions for the defence, preservation, inspection, recovery and control of the quality of life and the environment;
21	Drawing up legislation on the use of groundwater, establishing control and inspection measures;
22	Preserving the areas of the municipality that are part of the APA-Sul
23	Prioritise environmental education through the media, by implementing projects and activities in places of education, work, housing and leisure;
24	Manage and treat the solid waste generated by the municipality, including promoting educational campaigns and public policies aimed at contributing to the reuse, reduction, reutilisation and recycling of this waste;
25	Demand the recovery of degraded areas and guarantee compensation for damage caused to the environment;
26	Create an environmental information system;
27	Encourage and support the participation of citizens and their representative bodies in environmental control actions, promoting the implementation of environmental education actions in governmental and non-governmental plans, programmes and projects;
28	Promoting acoustic comfort in the city through actions by the municipal government, in partnership with companies, non-governmental organisations and the community;
29	Expand the air quality monitoring network and encourage the use of alternative fuels in motor vehicles, especially taxis, official cars and those that provide services to the municipality, creating the green fleet;
30	To define, through its own regulations, guidelines for the implementation of parks, squares and other green areas in the city, encompassing aspects of occupation and preservation of the natural heritage of the land;
31	Draw up a master plan for the city's green areas and afforestation, characterising and mapping them;
32	Create incentive mechanisms that favour partnerships with the private sector in the implementation and maintenance of green areas;
33	Promote, in line with the municipality's housing policy, actions to rescue or recover invaded public green areas and actions to prevent future invasions;
34	Encourage and adopt, where possible, environmentally friendly alternative technologies in the actions carried out by the public and private sectors;

	Adopting aspects of the environmental dimension in urban developments, taking into account indicators of comfort and environmental sustainability in their design, as a way of improving the quality of life of the population;
35	
36	Promote an appropriate policy for the implementation of green areas in towns and favelas;
37	Require institutions and public service concessionaires to guard, guarantee the integrity, urban treatment, maintenance and conservation of the domain and service strips under their responsibility.

Table 4: Description of the environmental legislation mentioned in Table 2

Legislation	Description of the law/decree
Law[0] 6.314/93	Provides for the establishment, in the Municipality of Belo Horizonte, of Private Ecological Reserves, by owner designation.
Law[0] 7.166/96	Establishes rules and conditions for the subdivision, occupation and use of urban land in the municipality.
Law[0] 8.616/2003	It contains the Code of Postures of the Municipality of Belo Horizonte. Chapter VI, Article 221, states that the location of a dump must be indicated by the executive power.
Law no. [0] 10.522/201 2	Establishes the Sustainable Management System for Construction Waste and Voluminous Waste - SGRCC - and the Municipal Plan for Integrated Management of Construction Waste and Voluminous Waste - PMRCC, and makes other provisions.
Law no. [0] 4.998/88	Provides for the granting of licences and the control of radioactive material and radiation sources in the Municipality of Belo Horizonte and makes other provisions.
Law[0] 7.277/97	Institutes Environmental Licensing and makes other provisions.
Law no. 9.011/ 2005	Provides for *the* organisational structure of the Executive Branch's Direct Administration and makes other provisions.
Decree no. 14.370/201 1	Approves the Statute of the Municipal Parks Foundation and makes other provisions.
Decree no. 14.708/201 1	Establishes rules and procedures for partnerships between the Municipality of Belo Horizonte and society regarding the adoption of public green areas - Adopt the Green Programme - and makes other provisions.

Law[0] 9.505/2008	Provides for the control of noise, sounds and vibrations in Belo Horizonte.
Law No. 8.262/2001	Provides for air monitoring and control in the municipality. Operação Oxigênio (Operation Oxygen) - monitoring of air pollution caused by diesel vehicles, created in 1988 through an agreement between Belo Horizonte City Hall and the Minas Gerais State Government.
Law[0] 10.175/2011.	Establishes the Municipal Policy for Mitigating the Effects of Climate Change.
Decree no. 14.794/201 2	Promotes the Municipal Policy for Mitigating the Effects of Climate Change through the Municipal Plan for Reducing Greenhouse Gas Emissions.
Decree no. 14.060/2010	Code of postures of the municipality of Belo Horizonte.
Decree no. 15.745/2014	Creates the Steering Committee, the Advisory Board and the Executive Secretariat for the preparation of the Municipal Plan for Integrated Solid Waste Management - PMGIRS of the Municipality of Belo Horizonte and makes other provisions.
Decree No. 5.893 of 16 March 1988	Regulates Law No.[0] 4.253, of 4 December 1985, which provides for the policy of protection, control and conservation of the environment and improvement of the quality of life in the Municipality of Belo Horizonte.
Law no. [0] 9.505/08	Provides for the control of noise, sounds and vibrations in the Municipality of Belo Horizonte and makes other provisions.
Law[0] 4.495/1986	Regulates the release into *the* atmosphere of exhaust gases from metropolitan public buses in use in the Municipality of Belo Horizonte.
Law[0] 9.340/2007	Establishes the Belo Horizonte Biodiesel Programme.
Law No. 10.175	Establishes the Municipal Policy for Mitigating the Effects of Climate Change.
Law no. [0] 6.314/93	Provides for the establishment, in the Municipality of Belo Horizonte, of Private Ecological Reserves, by owner designation.
Decree No. 5.893 of 16 March 1988	Regulates Law No. 4.253, of 4 December 1985, which provides for the policy of protection, control and conservation of the environment and improvement of the quality of life in the Municipality of Belo Horizonte.

With regard to the laws and decrees mentioned in Table 2, many of them also establish public policies and/or programmes implemented in the municipality. What's more, it can be seen that the same programmes address issues dealt with in many directives, and are related to these.

During the search for legislation, public policies and environmental programmes implemented in Belo Horizonte, it was noted that some of the websites of municipal bodies that said there was some legislation related to the subject, which was addressed in some guideline, this legislation was not found, such as the legislation on pruning, cutting, planting and transplanting trees in the capital. Alternatively, some bills were found that were also related to the guidelines, but it was not possible to find information about the sanctioning of these bills on the websites of the city's public bodies.

Another factor was that a lot of data could not be found, either in books, articles or on the websites of municipal and other bodies, about what was covered in some of the guidelines. The information that couldn't be found and other questions that arose during the research were sent to the city council in a letter so that they could be answered by a public authority involved in environmental issues in Belo Horizonte. However, until the research was finalised, the authors had received no response from the municipal body. This hindered the progress of the research and made it difficult or impossible to conclude on many of the issues addressed in this study.

Based on a qualitative study of the 37 environmental guidelines in the Master Plan, it can be seen that their content demonstrates the great relevance given to the environmental issue in Belo Horizonte, covering various aspects, especially in relation to green areas and environmental control intrinsic to the municipality. However, aspects related to sustainable technologies and water resources could be better addressed in the Plan.

After a qualitative analysis of the environmental legislation and public policies implemented in the municipality, it can be seen that most of them include various environmental guidelines and reveal their essential importance in maintaining and overseeing environmental issues in the capital of Minas Gerais, at least in theory. In addition, many of the public policies seek to involve the participation of society and make citizens come closer and realise the true significance of the environment for the municipality.

It is important to point out that the aim of the research was to analyse the environmental issue in Belo Horizonte's Master Plan, but based on studies of legislation and public policies implemented in the city, according to information available on the websites of municipal public bodies, such as the City Council and the Mayor's Office, and other sources, such as articles and other academic works. Visits to parks or evidence that what was theorised was being implemented, according to the agencies, was not possible due to the lack of time to carry out this activity during the research.

CHAPTER 5

FINAL CONSIDERATIONS

From the research it was possible to highlight some aspects that deserve consideration:

1. The bibliographical research related to environmental issues and urban planning, and the qualitative analysis of municipal legislation related to the environment favoured an understanding of how environmental and urban planning is carried out in Belo Horizonte.

2. Belo Horizonte's Master Plan has a predominance of environmental guidelines for green areas and environmental control instruments. Despite this, it was found that many of the environmental guidelines have not been implemented through public policies and legislation, thus demonstrating that the environmental issue is not fully addressed in municipal planning.

3. From the Master Plan for the capital of Minas Gerais, it can be seen that the government's vision of environmental issues is closely linked to a physical-ecological dimension only, dissociated from the social dimension that is intimately related to urban environmental problems.

4. Environmental guidelines must be implemented through various instruments, such as legislation, technical standards and public policies related to environmental issues. However, by analysing these instruments at municipal level, it was possible to see that approximately 40% of environmental guidelines have not been effectively implemented. There is therefore a need for the public authorities involved with the city's environmental issue to address the guidelines that have not yet been contemplated.

5. The delay in replying to the letter sent to City Hall, which hampered the progress

of the research and made it impossible to conclude on some of the issues addressed in the study.

6. Lack of information provided by municipal public bodies regarding the content of some environmental guidelines and the enactment of some laws.

7. Little time was available, which made it impossible to carry out activities involving visits to municipal green areas to see in practice how public policies and laws applied in the city work.

CHAPTER 6

ACKNOWLEDGEMENTS

The authors would like to thank the public bodies that made the research possible: the Federal Technological Education Centre of Minas Gerais (CEFET-MG), which funded the author's studies through its Junior Scientific Initiation Scholarship Programme (BIC-JR), together with the Minas Gerais State Research Support Foundation (FAPEMIG) and the National Council for Scientific and Technological Development (CNPq).

CHAPTER 7

REFERENCES

ABRANCHES, Mônica. Urban planning in Belo Horizonte: analysing the role of municipal councils in city management. **Cadernos Metrópole**, São Paulo, v. 11, n. 22, p. 495-517, jul./dez. 2009.

BAHIA, Cláudio Listher Marques. Belo Horizonte: a city for modernity in Minas Gerais. **Cadernos de Arquitetura e Urbanismo**, Belo Horizonte, v. 12, n. 13, p. 185200, dez. 2005. Disponívelem : <http://www2.pucminas.br/imagedb/documento/DOC_DSC_NOME_ARQUI200705 1409 1138.pdf>. Accessed on: 23 January 2016.

BELO HORIZONTE. City Council. **Masterplan**. Belo Horizonte, 2016. Available at:<http://www.cmbh.mg.gov.br/atividade-legislativa/planejamento-orcamento-publico/plano_diretor>. Accessed on: 25 March 2016.

BELO HORIZONTE. City Hall. Law No. 7.165: of 27 August 1996. Establishes the Master Plan for the Municipality of Belo Horizonte. Belo Horizonte, 1996.

BELO HORIZONTE. City Hall. Law No. 7.166: of 27 August 1996. Establishes rules and conditions for the subdivision, occupation and use of urban land in the municipality. Belo Horizonte, 1996.

BELO HORIZONTE. City Hall. **Environment**. Belo Horizonte, 2016. Available at: <http://portalpbh.pbh.gov.br/pbh/ecp/comunidade.do?app=meioambiente>. Accessed on: 27 March 2016.

BELO HORIZONTE. Belo Horizonte Metropolitan Region Planning Superintendence. **The development process of Belo Horizonte**. Belo Horizonte: Plambel, 1979.

BRAZIL. Law no. 10.257 of 10 July 2001. Regulates articles 182 and 183 of the

Federal Constitution, establishes general guidelines for urban policy and makes other provisions. **Official Gazette of the Federative Republic of Brazil, Legislative Branch**, Brasília, DF, 11 July 2001. Available at: <http://www.planalto.gov.br/ccivil_03/leis/LEIS_2001/L10257.htm>. Accessed on: 7 January 2016.

CARLOS, Ana Fani Alessandri. **The city**. 9. ed., 1ª reimpr. São Paulo: Contexto, 2013.

CORRÊA, Roberto Lobato. **The Urban Space**. 4 ed. São Paulo: Ática, 2005.

COSTA, Ana Carolina Silva da; ARGUELHES, Delmo de Oliveira. Social hygienisation through urban planning in Belo Horizonte in the early years of the 20th century. **Univ. Hum.**, Brasília, v. 5, n. 1/2, p. 109-137, jan./dez. 2008. Available at: <http://www.publicacoesacademicas.uniceub.br/index.php/universitashumanas/article /dow nload/878/612>. Accessed on: 22 January 2016.

COSTA, Camila Rodrigues. **Planning, public action and property dynamics in the recent history of Belo Horizonte**. 2006. 26 p. School of Architecture, Federal University of Minas Gerais, Belo Horizonte. Available at: <http://www.arq.ufmg.br/lab- urb/wp-content/uploads/2013/10/Camila-Costa.pdf>. Accessed on: 23 January 2016.

DUARTE, Fábio. **Urban Planning**. 20 ed. Curitiba: Ibpex, 2007; 177 p.

SANTOS, Milton. [1996]. **The Nature of Space**: Technique and Time. Reason and Emotion. 4. Ed 7ª reimpr. São Paulo: Editora da Universidade de São Paulo, 2012. Available at: <http://www.nepam.unicamp.br/arqueologiapublica/artigos/artigo_graduacao_8.pdf>. Accessed on: 18 June 2015.

MELO, Denise M. Urban planning in Belo Horizonte: a study of the main planning proposals drawn up for the city. 1991. 143p. Monograph (Specialisation in Urban Planning) - School of Architecture, Federal University of Minas Gerais, Belo Horizonte.

PEIXOTO, N. B. (1996). **Urban landscapes**. São Paulo, Senac.

RODRIGUES, Rodrigo da Silva; FILHO, Francisco de Assis Veloso. The trajectory of urban planning in Brazil: an attempt at periodisation. In: ENCONTRO NACIONAL DA ASSOCIAÇÃO NACIONAL DE PÓS-GRADUAÇÃO E PESQUISA EM GEOGRAFIA, 10., 2013, UNICAMP, Campinas.

SILVA, José Afonso. **Brazilian urban law**. São Paulo: Malheiros, 1995.

VILLAÇA, Flávio. **Dilemmas of the Masterplan**. In: CEPAM. The municipality in the 21st century: scenarios and perspectives. São Paulo: Fundação Prefeito Faria Lima - Cepam, 1999. p. 237 - 247.

Printed by Books on Demand GmbH, Norderstedt / Germany